We hope this book has been informative and helpful on your journey to understanding and celebrating older adults. Thank you for your interest and support!

Title: Evolution of Electric Vehicles: The Birth of a New Era

Subtitle: Innovations and Roadblocks

Series: Ride Through Time: The Story of World Vehicles

By M.J. Knightly

"The transportation industry is on the brink of a revolution, with advancements in autonomous vehicles, electric cars, and new modes of transportation like hyperloops and vertical takeoff and landing aircrafts."
Mary Barra, CEO of General Motors

"Innovation in transportation is not only about speed and efficiency, but also about sustainability and reducing our impact on the environment."
Patricia Espinosa, Executive Secretary of the United Nations Framework Convention on Climate Change

"The future of transportation will be about creating a seamless and integrated experience, where different modes of transportation work together to get people where they need to go."
Dara Khosrowshahi, CEO of Uber

"The age of electric vehicles is here, and it's exciting to see how technology is driving the transformation of our transportation systems."
Elon Musk, CEO of Tesla and SpaceX

"Transportation is not just about getting from point A to point B, it's about the freedom and opportunities that come with mobility. Innovations in transportation are critical for empowering individuals and communities around the world."
Ban Ki-moon, former Secretary-General of the United Nations

"As transportation continues to evolve, we have the opportunity to create a more equitable and accessible system that benefits all people, regardless of their background or circumstances."
Keisha Lance Bottoms, former mayor of Atlanta, Georgia

"Innovation is key to unlocking the potential of transportation, and we need to encourage and invest in new ideas and technologies that will shape the future of mobility."
Hiroto Saikawa, former CEO of Nissan Motor Company

Table of Contents

Introduction
The need for sustainable transportation

As concerns about climate change and pollution continue to grow, sustainable transportation has become a crucial issue in today's society. The need to reduce greenhouse gas emissions and air pollution has led to the development of alternative fuel vehicles, including electric vehicles (EVs), which are powered by electricity stored in batteries.

The Need for Sustainable Transportation

The transportation sector is one of the largest sources of greenhouse gas emissions, accounting for approximately 29% of total US greenhouse gas emissions. The majority of these emissions come from cars and light-duty trucks, which are powered by gasoline and diesel fuel. These emissions contribute to climate change, which has numerous negative impacts on the environment, including rising sea levels, more frequent and severe weather events, and the loss of biodiversity.

In addition to climate change, transportation also contributes to air pollution, which can have significant impacts on human health. Exposure to air pollution has been linked to respiratory and cardiovascular diseases, as well as premature death. According to the World Health

Organization (WHO), outdoor air pollution is responsible for approximately 4.2 million premature deaths each year.

Given these challenges, there is a growing need for sustainable transportation solutions that can help reduce emissions and improve air quality. Electric vehicles are one such solution. Unlike gasoline and diesel vehicles, EVs produce zero emissions at the tailpipe. This means that they can help reduce greenhouse gas emissions and improve air quality, particularly in urban areas where air pollution is a major concern.

Furthermore, EVs can also help reduce our dependence on oil, which is a finite and non-renewable resource. By using electricity from renewable sources, such as wind and solar power, we can create a more sustainable and resilient energy system.

However, the adoption of electric vehicles faces several challenges, including high costs, limited driving range, and the need for charging infrastructure. To overcome these challenges, governments, businesses, and individuals must work together to create supportive policies and invest in the necessary infrastructure.

Conclusion

In conclusion, the need for sustainable transportation has never been greater. The transportation sector is a significant contributor to greenhouse gas emissions and air pollution, which have negative impacts on both the environment and human health. Electric vehicles have the potential to reduce emissions and improve air quality, while also reducing our dependence on oil. However, overcoming the challenges to widespread adoption of EVs will require cooperation and investment from all sectors of society.

The early development of electric vehicles

Electric vehicles (EVs) have a long and fascinating history, dating back to the early days of the automobile. While the modern electric vehicle is often associated with the rise of Tesla and other contemporary companies, the history of electric vehicles spans over a century. In this chapter, we will explore the early development of electric vehicles, from their early beginnings to the early 20th century.

The Early Development of Electric Vehicles

The first electric vehicles appeared in the mid-19th century, around the same time as gasoline-powered vehicles. The first electric vehicle was invented by Scottish inventor Robert Anderson in 1832. However, it was not until the late 1800s that electric vehicles began to gain popularity, particularly in urban areas.

One of the earliest and most successful electric vehicle manufacturers was the Electric Vehicle Company, which was founded in 1897 in New York City. The company produced a range of electric vehicles, including taxis, buses, and private cars. Electric vehicles were particularly well-suited for urban environments, where their quiet operation and lack of emissions made them an attractive alternative to the noisy and polluting gasoline-powered vehicles of the time.

Despite their early success, electric vehicles faced several challenges in the early 20th century. One of the biggest challenges was the development of the internal combustion engine, which quickly became the dominant technology in the automotive industry. Gasoline and diesel-powered vehicles had several advantages over electric vehicles, including longer range and faster refueling times. Additionally, the discovery of vast oil reserves in the early 1900s led to a significant drop in the cost of gasoline, making gasoline-powered vehicles more affordable than their electric counterparts.

As a result, electric vehicles fell out of favor in the early 20th century, and by the 1930s, they were largely confined to niche markets, such as golf carts and milk delivery trucks.

Conclusion

In conclusion, the early development of electric vehicles was marked by innovation and experimentation. Electric vehicles were some of the earliest automobiles and were well-suited for urban environments due to their quiet operation and lack of emissions. However, the development of the internal combustion engine and the discovery of vast oil reserves led to the decline of electric vehicles in the early 20th century. Despite their initial setbacks, electric vehicles would experience a revival of interest in the latter part of the 20th century, as concerns about climate change and air pollution once again brought attention to the need for sustainable transportation.

The current state of the EV industry

The electric vehicle (EV) industry has seen remarkable growth in recent years, driven by a combination of factors, including advances in battery technology, government incentives, and increased public awareness of the need for sustainable transportation. In this chapter, we will explore the current state of the EV industry, including the major players, market trends, and the challenges and opportunities facing the industry.

The Current State of the EV Industry

The EV market has grown rapidly in recent years, with global EV sales increasing from just over 450,000 in 2015 to more than 3.2 million in 2020. This growth has been driven in part by government incentives, such as tax credits and subsidies, which have helped to make EVs more affordable for consumers.

The EV market is dominated by a few major players, including Tesla, which has become synonymous with the EV industry, as well as established automakers such as Volkswagen, General Motors, and Nissan. These companies have invested heavily in the development of EV technology, and as a result, they have been able to produce high-quality, reliable vehicles that are increasingly competitive with their gasoline-powered counterparts.

One of the most significant challenges facing the EV industry is the limited range and long charging times of many EVs. While the range of EVs has improved significantly in recent years, with some models now able to travel over 300 miles on a single charge, range anxiety remains a significant concern for many consumers. Additionally, the lack of a robust charging infrastructure is a major barrier to widespread adoption of EVs, particularly in rural areas.

Despite these challenges, the future of the EV industry looks bright. As battery technology continues to improve, EVs are becoming increasingly competitive with gasoline-powered vehicles, with some experts predicting that EVs could become cost-competitive as early as 2025. Additionally, governments around the world are taking steps to incentivize the transition to EVs, with many countries setting ambitious targets for the phase-out of gasoline-powered vehicles.

Conclusion

In conclusion, the current state of the EV industry is characterized by rapid growth, driven by advances in battery technology and government incentives. While the industry still faces significant challenges, such as range anxiety and the lack of a robust charging infrastructure, the future looks bright for EVs. As the cost of battery technology continues to fall and the range of EVs continues to improve, we can expect to see increasing numbers of consumers making the switch to electric. The transition to a fully electric transportation system will not happen overnight, but with the continued support of governments, automakers, and consumers, it is a transition that is well underway.

Chapter 1: The Early Days of Electric Vehicles
The first electric cars

Electric vehicles have been around for over a century, and while they have seen a surge in popularity in recent years, their history goes back much further. In this chapter, we will explore the early days of electric vehicles, including the first electric cars, their early development, and the challenges they faced in competing with gasoline-powered vehicles.

The First Electric Cars

The first electric cars were developed in the late 19th century, around the same time as the first gasoline-powered cars. In fact, electric vehicles were initially more popular than gasoline-powered vehicles, with electric cars outselling gasoline-powered cars in the US until the early 1910s.

One of the first electric cars was developed by Thomas Davenport in 1835, who invented the first electric motor. However, it wasn't until the 1880s that practical electric vehicles began to be developed. One of the pioneers in this field was British inventor Thomas Parker, who in 1884 built the first electric car with a rechargeable battery.

In the United States, the first successful electric car was built by William Morrison in 1891. His car, which was nicknamed the "horseless carriage," had a top speed of 14 miles per hour and could travel up to 50 miles on a single charge.

Another notable electric car from this period was the Baker Electric, which was produced by the Baker Motor Vehicle Company in Ohio from 1899 to 1916. The Baker Electric was known for its reliability and was popular with wealthy customers, including some members of the royal families of Europe.

Despite their early success, electric cars faced several challenges that would limit their popularity in the years to come.

Limitations of Early Electric Cars

One of the biggest challenges facing early electric cars was their limited range. Early electric vehicles had a range of around 50 miles on a single charge, which made them unsuitable for long-distance travel. Additionally, the batteries used in these vehicles were heavy and bulky, making them difficult to transport and limiting the space available for passengers and cargo.

Another challenge facing electric cars was their relatively low top speed. While some early electric vehicles could reach speeds of up to 25 miles per hour, this was still slower than gasoline-powered cars, which could reach speeds of up to 60 miles per hour.

Finally, the cost of electric vehicles was a significant barrier to their widespread adoption. While electric cars were initially cheaper than gasoline-powered cars, as the technology developed, they became increasingly expensive. Additionally, the batteries used in electric cars were expensive to produce and had a relatively short lifespan, which meant that they needed to be replaced frequently.

Despite these challenges, electric cars remained popular with some customers, particularly in urban areas where their quiet operation and lack of emissions made them attractive.

Conclusion

In conclusion, the early days of electric vehicles were characterized by innovation and experimentation, as inventors and engineers worked to develop a practical electric car. While early electric vehicles faced several challenges, including limited range and high costs, they also had several advantages, such as their quiet operation and lack of emissions.

As we will see in the next chapter, the rise of gasoline-powered cars in the early 20th century would pose a significant challenge to the electric vehicle industry, leading to a decline in their popularity that would last for several decades. However, the potential of electric cars would not be forgotten, and as we will see later in this book, they would eventually make a comeback, driven by advances in battery technology and a growing awareness of the need for sustainable transportation.

While early electric vehicles were an exciting new technology, they also had some significant limitations that prevented them from achieving widespread use. Some of the most notable limitations included:

1. Limited range: Early electric cars had a very limited range, usually less than 50 miles on a single charge. This made them impractical for long-distance travel, and even short trips could be challenging if the driver needed to go too far from home.

2. Slow speed: Early electric cars were also relatively slow compared to gasoline-powered vehicles. This was partly due to the limitations of battery technology at the time, but it also had to do with the design of the vehicles themselves. Electric cars typically had fewer gears and a less powerful motor than gasoline cars, which made them slower and less responsive.

3. Long charging times: Charging an early electric car could take many hours, making it difficult to use the vehicle for multiple trips in a day. This was partly due to the limitations of battery technology, but it was also because early charging systems were not very efficient.

4. High cost: Early electric cars were expensive to produce and purchase, partly because the technology was

still new and partly because they required specialized components like batteries and electric motors. This made them a luxury item that only a small number of people could afford.

5. Limited infrastructure: Early electric cars were also limited by the lack of infrastructure to support them. There were few charging stations or repair shops available, which made it difficult for drivers to keep their vehicles running smoothly.

Despite these limitations, early electric cars were an important step forward in the development of sustainable transportation. They laid the foundation for future innovations in battery technology and electric vehicle design, which would eventually overcome many of these early challenges.

The impact of the internal combustion engine

The internal combustion engine had a profound impact on the development of transportation in the 20th century, and its emergence had both positive and negative effects on the growth of electric vehicles. In this section, we will explore the impact of the internal combustion engine on the early development of electric vehicles.

The internal combustion engine was first developed in the mid-19th century and quickly became the dominant technology for powering vehicles. Gasoline engines were more powerful and had a longer range than early electric vehicles, which made them more practical for many applications. They were also more versatile, as they could be used in a variety of vehicles, from cars and trucks to boats and airplanes.

As the use of gasoline engines grew, electric vehicles became less popular. Gasoline engines were cheaper to produce and maintain, and gasoline was more widely available than electricity. This made gasoline vehicles more practical for most consumers, and electric vehicles were relegated to niche applications, such as delivery trucks and golf carts.

The rise of the internal combustion engine also had negative environmental impacts. Gasoline engines produce pollutants like carbon dioxide, nitrogen oxides, and particulate matter, which contribute to air pollution and climate change. The widespread use of gasoline engines has also contributed to the depletion of fossil fuels, which are finite resources that are becoming increasingly scarce.

In recent years, concerns about the environmental impact of gasoline engines have led to renewed interest in electric vehicles. Modern electric vehicles are much more practical and affordable than their early counterparts, thanks to advances in battery technology and improvements in electric vehicle design. They offer a clean and sustainable alternative to gasoline vehicles, and they are becoming increasingly popular with consumers.

In conclusion, the internal combustion engine had a significant impact on the development of transportation in the 20th century. While it played an important role in the growth of gasoline vehicles, it also contributed to negative environmental impacts and hindered the growth of electric vehicles. However, the emergence of modern electric vehicles has the potential to revolutionize the transportation industry and reduce our dependence on fossil fuels.

Chapter 2: The Rise and Fall of Electric Vehicles
The growth of EVs in the early 20th century

In the early 20th century, electric vehicles saw a significant surge in popularity, driven by advancements in technology and changing societal attitudes towards transportation.

One of the most significant early adopters of electric vehicles was the taxi industry. In major cities such as New York, electric taxis were a common sight on the streets. In fact, by the early 1910s, there were more electric taxis in New York City than there were gasoline-powered ones. The reasons for this were twofold. First, electric taxis were much quieter than their gasoline-powered counterparts, making them a popular choice for city-dwellers who were looking for a more peaceful mode of transportation. Second, electric taxis were much easier to operate than gasoline-powered ones, which required a significant amount of manual labor to start and maintain.

Beyond the taxi industry, electric vehicles also gained popularity among private consumers. One notable example was the Detroit Electric, which was produced from 1907 to 1939. The Detroit Electric was a stylish and reliable electric car that was marketed primarily to women. It was one of the most successful electric cars of its time, with more than 13,000 units sold.

Electric delivery trucks also saw a surge in popularity during this time period. Companies such as General Electric and Westinghouse developed electric trucks that were used for a variety of purposes, including mail delivery and milk delivery.

The growth of electric vehicles in the early 20th century was driven in part by advancements in battery technology. Lead-acid batteries, which were first developed in the mid-19th century, became smaller, more powerful, and more affordable over time, making them a viable option for electric vehicles.

However, the growth of the electric vehicle industry was short-lived. The invention of the electric starter motor in 1912 made gasoline-powered cars much easier to operate, eliminating one of the major advantages that electric cars had. Additionally, the discovery of vast oil reserves in Texas in the early 20th century led to a significant drop in the price of gasoline, making gasoline-powered cars much more affordable than electric ones.

By the 1920s, the electric vehicle industry was in decline. Many electric vehicle manufacturers went out of business, and the remaining ones focused primarily on producing electric golf carts and other low-speed vehicles. It would be several decades before the electric vehicle industry would see a significant resurgence.

The decline of EVs due to cheap gasoline and oil

The early 20th century saw a rapid growth of electric vehicles (EVs) in the United States, Europe, and other parts of the world. However, the industry was short-lived, and by the 1920s, EVs had largely disappeared from the roads. One of the main reasons for this decline was the rise of gasoline-powered vehicles, which were cheaper, had longer range, and were more convenient to use. This chapter will explore the decline of EVs due to cheap gasoline and oil.

The first gasoline-powered vehicles were introduced in the late 19th century, but it was not until the early 20th century that they became popular. Gasoline vehicles were initially expensive, and their engines were noisy, dirty, and unreliable. In contrast, electric vehicles were silent, clean, and easy to operate. They were popular with women drivers because they did not require manual cranking to start the engine, which was necessary for gasoline vehicles at the time.

However, the main advantage of gasoline vehicles over electric vehicles was their longer range. Gasoline vehicles could travel much farther than electric vehicles, which were limited by the capacity of their batteries. This meant that gasoline vehicles were more versatile and could be used for longer trips, which made them more attractive to consumers.

Another factor that contributed to the decline of EVs was the decreasing cost of gasoline. In the early 20th century, gasoline was cheap and plentiful, thanks to new discoveries of oil in the United States and other countries. As the cost of gasoline decreased, the cost of operating gasoline vehicles also decreased, making them more affordable for consumers.

In addition, the development of the electric starter for gasoline vehicles in 1912 made gasoline vehicles much more convenient to use. Before the electric starter, gasoline vehicles had to be manually cranked to start the engine, which was difficult and dangerous, especially for women and elderly drivers. The electric starter made gasoline vehicles much easier to operate, which made them more appealing to consumers.

The decline of EVs was also due to the lack of investment in EV technology. With the rise of gasoline vehicles, many investors turned their attention to the gasoline industry, which was more profitable. This led to a decline in research and development of EV technology, which meant that EVs did not improve as quickly as gasoline vehicles.

Furthermore, the decline of EVs was also due to the lack of infrastructure for charging and servicing electric vehicles. Gasoline vehicles could be refueled at gas stations, which were ubiquitous, but electric vehicles had few charging stations, which made them inconvenient for long-distance travel. This lack of infrastructure made it difficult for EVs to compete with gasoline vehicles.

In conclusion, the decline of electric vehicles in the early 20th century was due to a combination of factors, including the rise of gasoline vehicles, the decreasing cost of gasoline, the lack of investment in EV technology, and the lack of infrastructure for charging and servicing electric vehicles. While electric vehicles had many advantages over gasoline vehicles, they were not able to compete with gasoline vehicles on price, convenience, and range.

The revival of interest in EVs in the 1990s

The 1990s marked a turning point in the history of electric vehicles, as interest in electric cars began to surge once again. This resurgence was driven by a number of factors, including concerns about air pollution, global warming, and the long-term sustainability of fossil fuels. Here, we will explore the revival of interest in EVs in the 1990s and the key developments that helped to pave the way for the modern electric vehicle industry.

One of the key drivers behind the renewed interest in electric vehicles in the 1990s was the California Air Resources Board (CARB), which introduced a mandate requiring automakers to produce zero-emission vehicles (ZEVs) as a way of combatting air pollution in the state. This mandate required automakers to produce a certain number of ZEVs as a percentage of their total sales, with the percentage increasing over time.

While some automakers initially resisted the CARB mandate, others saw it as an opportunity to invest in electric vehicle technology and gain a competitive edge in the emerging market. One of the first automakers to invest heavily in electric vehicles was General Motors, which developed the EV1, an all-electric car that was first introduced in 1996.

The EV1 was a significant departure from earlier electric vehicles, which were typically small and underpowered. The EV1 was designed to be a practical and efficient car, with a range of up to 140 miles on a single charge. While the EV1 was initially available only for lease, it quickly gained a devoted following among early adopters of electric vehicle technology.

Other automakers also began to introduce electric vehicles during this time, including Ford, which developed the Ranger EV, and Honda, which introduced the EV Plus. While these early electric vehicles were still relatively expensive and limited in terms of range, they helped to build momentum for the emerging electric vehicle industry.

Another key development during this time was the introduction of nickel-metal hydride (NiMH) batteries, which offered a significant improvement in energy density over the lead-acid batteries used in earlier electric vehicles. NiMH batteries were more efficient, longer-lasting, and more environmentally friendly than lead-acid batteries, and helped to make electric vehicles more practical for everyday use.

The revival of interest in electric vehicles in the 1990s also paved the way for the development of hybrid vehicles, which combine an internal combustion engine with an electric motor and battery pack. The Toyota Prius, which was first introduced in Japan in 1997 and in the United States in 2000, quickly became a symbol of the emerging hybrid vehicle market.

Overall, the 1990s marked a period of significant innovation and growth in the electric vehicle industry, as automakers and policymakers alike recognized the potential of electric vehicle technology to address pressing environmental and sustainability challenges. While the industry still faced many obstacles and uncertainties, the groundwork was laid for the development of the modern electric vehicle industry that we see today.

Chapter 3: The Birth of Modern Electric Vehicles
The development of the lithium-ion battery

The development of the lithium-ion battery was a crucial turning point for modern electric vehicles. Before the invention of lithium-ion batteries, electric cars had limited range and were not suitable for long-distance travel. This section will explore the history of the lithium-ion battery and its impact on the electric vehicle industry.

History of Lithium-ion Batteries The idea of using lithium for batteries can be traced back to the 1970s when researchers discovered that lithium could be used to create a high-energy-density battery. However, it was not until the 1980s that a working prototype was developed by Akira Yoshino, who is now credited as the inventor of the lithium-ion battery.

The first commercial lithium-ion battery was introduced in 1991 by Sony Corporation. It was smaller, lighter, and had a higher energy density than previous battery technologies. This made it ideal for use in portable electronic devices like laptops and mobile phones.

Impact on Electric Vehicles The development of the lithium-ion battery was a game-changer for the electric vehicle industry. It allowed electric vehicles to have a longer range and reduced the need for frequent recharging.

Lithium-ion batteries are also more durable and have a longer lifespan than other battery technologies.

The emergence of Tesla and the Roadster in 2008 was a key moment for the lithium-ion battery. Tesla's Roadster was the first all-electric vehicle to use lithium-ion batteries, and it had a range of over 200 miles on a single charge. This was a major milestone for the electric vehicle industry and helped to increase interest in electric cars.

Lithium-ion batteries have continued to improve in terms of performance and cost. As a result, electric vehicles have become more affordable and practical for everyday use. In recent years, the cost of lithium-ion batteries has decreased significantly, making electric vehicles more competitive with traditional gasoline-powered cars.

Future Developments There is ongoing research to improve the performance and reduce the cost of lithium-ion batteries. One area of focus is developing solid-state batteries, which have the potential to provide higher energy density and faster charging times than current lithium-ion batteries.

There is also increasing interest in battery recycling and reuse. As the demand for electric vehicles continues to grow, it will become increasingly important to find sustainable ways to manufacture and dispose of lithium-ion batteries.

Conclusion The development of the lithium-ion battery has had a profound impact on the electric vehicle industry. It has enabled electric vehicles to have a longer range and be more practical for everyday use. As battery technology continues to improve, electric vehicles are likely to become even more affordable and practical, driving the transition to a cleaner, more sustainable transportation system.

The emergence of Tesla and the Roadster

While several automakers, including Nissan and Chevrolet, had been producing electric vehicles since the early 2000s, it was the emergence of Tesla that truly sparked a renewed interest in electric cars. Founded in 2003 by Elon Musk, Tesla aimed to create a high-performance electric car that could compete with the best gasoline-powered sports cars.

The company's first product, the Tesla Roadster, was introduced in 2008. It was a high-end, two-seat sports car that boasted impressive acceleration and range. The Roadster was based on a Lotus Elise chassis, but was fitted with a powerful electric motor and an advanced lithium-ion battery pack.

The Roadster's performance was nothing short of impressive. It could accelerate from 0 to 60 miles per hour in just 3.7 seconds and had a top speed of 125 miles per hour. It also had a range of 245 miles on a single charge, which was a significant improvement over previous electric cars.

Despite its high price tag of over $100,000, the Roadster quickly gained a following among early adopters and enthusiasts. It was seen as a symbol of a new era of electric cars that could be both stylish and high-performance.

In addition to the Roadster, Tesla also developed a proprietary charging infrastructure to support its cars. The company installed "Supercharger" stations along major highways in the US and Europe, allowing Tesla owners to quickly charge their cars on long trips.

The success of the Roadster and Tesla's unique approach to electric cars helped the company attract significant investment and develop a strong brand identity. It also paved the way for Tesla's future success with the Model S, Model X, and other electric vehicles.

The introduction of the Nissan Leaf and Chevy Volt

The introduction of the Nissan Leaf and Chevy Volt marked a turning point in the history of modern electric vehicles. These two vehicles demonstrated that EVs could be practical for everyday use and not just limited to niche applications.

In 2010, Nissan launched the Leaf, which was the world's first mass-produced electric vehicle. The Leaf had a range of approximately 100 miles on a single charge, making it suitable for most daily commutes. The Leaf was also equipped with a range of features that made it a practical car for everyday use, such as air conditioning and heating, power windows and locks, and a navigation system.

The Chevy Volt was also launched in 2010 and was a hybrid electric vehicle that had an electric range of approximately 35 miles before switching to a gasoline-powered generator to extend its range. The Volt was marketed as a car that could go "all-electric" for most daily commutes while still providing the range and flexibility of a gasoline-powered car.

The introduction of these two vehicles helped to change public perception of electric vehicles and showed that they could be practical for everyday use. They also helped to spur competition among automakers and encourage the development of new EV technologies.

Since the launch of the Leaf and Volt, both vehicles have undergone several updates and improvements. The Leaf now has a range of up to 226 miles, while the Volt has been replaced by the Chevy Bolt, which has a range of over 250 miles on a single charge.

Other automakers have also entered the EV market, with companies like Tesla, BMW, and Volkswagen launching a range of electric vehicles. These vehicles have helped to further increase the popularity and acceptance of electric vehicles and have driven down the cost of EV technology.

Overall, the introduction of the Nissan Leaf and Chevy Volt marked a significant milestone in the history of electric vehicles. These vehicles helped to change public perception of EVs and demonstrated that they could be practical for everyday use. The success of these vehicles also helped to spur competition and innovation in the EV industry, leading to the development of new technologies and the introduction of more practical and affordable electric vehicles.

Chapter 4: The Current State of Electric Vehicles
The growth of the EV market

The growth of the EV market has been significant in recent years, with a steady increase in sales and production worldwide. According to the International Energy Agency (IEA), in 2020, the global stock of electric passenger cars passed 10 million, which is a 43% increase from 2019. The growth of the EV market can be attributed to several factors, including government policies, advancements in technology, and changing consumer preferences.

One of the most significant drivers of EV adoption has been government policies aimed at reducing greenhouse gas emissions and improving air quality. Many countries have implemented incentives such as tax credits, rebates, and subsidies for the purchase of EVs, as well as regulations mandating automakers to produce a certain percentage of zero-emission vehicles. For instance, in the United States, the federal government offers a tax credit of up to $7,500 for the purchase of an EV. In addition, several states offer additional incentives, such as rebates and HOV lane access, to encourage EV adoption.

Another factor driving the growth of the EV market is advancements in technology, particularly in battery technology. Battery costs have been declining steadily, making EVs more affordable and reducing range anxiety. Moreover, the energy density of batteries has been improving, resulting in longer driving ranges for EVs. For instance, the Tesla Model S Long Range Plus can travel up to 402 miles on a single charge, while the Chevrolet Bolt EV has a range of 259 miles.

Changing consumer preferences have also played a role in the growth of the EV market. Consumers are becoming increasingly aware of the environmental impact of their transportation choices and are looking for alternatives to traditional gasoline-powered vehicles. Moreover, as the technology improves, EVs are becoming more appealing to consumers, with many models offering high performance, luxury features, and advanced technology.

The growth of the EV market is expected to continue in the coming years, with many automakers planning to introduce new EV models and increase their production capacity. For instance, Volkswagen plans to produce 1.5 million EVs per year by 2025, while General Motors aims to introduce 30 new EV models by 2025. In addition, several countries, including the United Kingdom and Norway, have set targets to phase out the sale of new gasoline and diesel-powered vehicles in the coming decades, which is likely to further accelerate the growth of the EV market.

However, challenges remain in the EV market, including the need for additional charging infrastructure and concerns about the environmental impact of battery production and disposal. Nevertheless, the growth of the EV market in recent years has demonstrated the potential of electric vehicles to transform the transportation sector and reduce greenhouse gas emissions.

The range and charging challenges

While electric vehicles (EVs) have come a long way since their early days, there are still some significant challenges that need to be addressed before they can become the dominant form of transportation. One of the biggest challenges facing EVs is range anxiety. The limited range of EVs and the time it takes to charge them can make them impractical for some drivers, particularly those who drive long distances.

Range anxiety is a term used to describe the fear that an EV driver experiences when they are concerned that their battery will run out of charge before they reach their destination. Even though the range of EVs has improved over the years, most EVs still have a maximum range of 200-300 miles per charge, which is much less than the range of most gasoline-powered cars. Additionally, the actual range of an EV can vary depending on driving conditions, weather, and how aggressively the driver accelerates.

To overcome range anxiety, there are a few solutions that have been proposed. One solution is to increase the range of EVs. Car manufacturers are working on developing new batteries that can store more energy, which would increase the range of EVs. Another solution is to improve charging infrastructure, which would allow drivers to charge their cars more quickly and easily, making it more convenient to own an EV.

Charging infrastructure is another challenge facing the EV industry. While there are now more charging stations than ever before, there are still not enough to meet the needs of all EV drivers. In addition, many charging stations are slow and require several hours to fully charge an EV. This can be a problem for drivers who need to charge their car quickly and get back on the road.

To address this issue, several companies are working on developing faster charging stations. For example, Tesla has developed its Supercharger network, which can charge a car in 30 minutes or less. Other companies, such as Electrify America, are also investing in faster charging infrastructure.

In addition to faster charging stations, there are also new technologies being developed that could make charging more convenient. For example, wireless charging technology is being developed that would allow drivers to charge their cars without having to plug them in. This technology could make charging as easy as parking in the right spot.

Another challenge facing EVs is the availability of charging stations in certain areas. While many cities and urban areas have a good number of charging stations, rural areas often have few or none. This can make it difficult for

EV drivers to take long trips, as they may not be able to find a charging station when they need one.

To address this issue, governments and private companies are investing in building more charging stations in rural areas. In addition, some companies are developing portable charging solutions that can be used in remote areas where there are no charging stations.

Conclusion

While there are still some challenges facing the EV industry, significant progress has been made in recent years. The range of EVs has improved, and charging infrastructure is becoming more widespread and convenient. As more car manufacturers introduce new EV models, and as governments and private companies invest in the infrastructure needed to support them, the adoption of EVs is likely to continue to grow.

The competition among EV manufacturers

The electric vehicle (EV) market has grown tremendously in recent years, and as a result, competition among EV manufacturers has become increasingly intense. This chapter will explore the current state of the EV market and the competition among manufacturers.

One of the biggest players in the EV market is Tesla. Since its founding in 2003, the company has become synonymous with EVs, with its vehicles often referred to as the gold standard in the industry. The Model S was a game-changer when it was introduced in 2012, and subsequent models, including the Model 3 and Model Y, have been highly successful.

Other established automakers have also entered the EV market, including General Motors (GM), Nissan, and Volkswagen (VW). GM's Chevrolet Bolt was one of the first affordable long-range EVs on the market, and the company is now investing heavily in its Ultium battery technology. Nissan's Leaf, which debuted in 2010, was an early player in the market, and the company continues to improve the vehicle with each new iteration. VW has also made significant investments in the EV market, with plans to launch several new models over the next few years.

Chinese automakers have also been making a splash in the EV market. One of the biggest players is BYD, which is backed by Warren Buffett's Berkshire Hathaway. The company has been making EVs since 2003 and is now the world's largest EV manufacturer. Other Chinese manufacturers, including NIO and Xpeng, have also gained traction in the market.

In addition to established automakers, there are also several up-and-coming EV manufacturers. Rivian, for example, has raised billions of dollars in funding and is set to launch its highly anticipated R1T electric pickup truck in 2022. Lucid Motors is another company to watch, with plans to launch its luxurious Air sedan later this year.

While Tesla remains the dominant player in the market, competition is fierce, and manufacturers are constantly pushing the boundaries of what is possible with EVs. Range and charging times, once major obstacles for the industry, have improved significantly in recent years, and new technologies are continuing to be developed to improve them further.

Manufacturers are also competing on price, with many offering lower-priced EVs to make them more accessible to a wider range of consumers. Government incentives and tax credits are also helping to make EVs more affordable, and many countries, including the United States and China, have set ambitious targets for electrification in the coming years.

In addition to competition among manufacturers, there is also competition among charging networks. Companies such as ChargePoint, EVgo, and Electrify America are vying for market share, and new players are entering the market all the time. The availability of charging infrastructure is crucial to the growth of the EV market, and manufacturers are working closely with charging companies to ensure that their vehicles are compatible with a range of charging stations.

Overall, the competition among EV manufacturers is driving innovation and pushing the industry forward. As the market continues to grow and mature, it will be interesting to see which companies emerge as leaders and how they will shape the future of transportation.

Chapter 5: The Impact of Electric Vehicles
The environmental benefits of EVs

Introduction: Electric vehicles (EVs) are becoming an increasingly popular mode of transportation due to their environmental benefits. They produce fewer emissions and pollutants compared to gasoline-powered cars, resulting in a significant reduction in air pollution and greenhouse gas emissions. In this chapter, we will discuss the environmental benefits of EVs and their impact on the environment.

Reducing Greenhouse Gas Emissions: One of the most significant environmental benefits of EVs is their ability to reduce greenhouse gas emissions. EVs do not emit any carbon dioxide (CO_2) or other pollutants from their tailpipes. Instead, they rely on electricity produced by power plants, which can be generated from renewable energy sources such as wind, solar, or hydropower.

A study by the Union of Concerned Scientists found that EVs produce less than half the CO_2 emissions of gasoline-powered cars, even when taking into account the emissions generated during the production of batteries and electricity. Furthermore, as the electricity grid continues to shift toward renewable energy sources, the emissions associated with EVs will continue to decrease.

Reducing Air Pollution: In addition to reducing greenhouse gas emissions, EVs also reduce air pollution. According to the Environmental Protection Agency (EPA), transportation is the largest contributor to air pollution in the United States, with cars and trucks responsible for over half of the total emissions of nitrogen oxides (NOx) and particulate matter (PM).

EVs produce no tailpipe emissions, which means they do not release any NOx, PM, or other pollutants into the air. This results in improved air quality, particularly in urban areas where air pollution can have a significant impact on public health.

Reducing Noise Pollution: EVs are also significantly quieter than gasoline-powered cars. The absence of an internal combustion engine means that there is no loud engine noise or exhaust noise, which can contribute to noise pollution in urban areas. This can lead to a more peaceful and enjoyable driving experience, especially in dense city centers.

Conclusion: The environmental benefits of EVs are significant and can have a positive impact on the environment, public health, and quality of life. By reducing greenhouse gas emissions, air pollution, and noise pollution, EVs are becoming an increasingly important part of the transportation sector. As the technology continues to improve and become more affordable, the environmental benefits of EVs will only continue to grow, making them an essential tool in the fight against climate change.

The potential to reduce dependence on oil

The widespread use of petroleum products has resulted in an increase in greenhouse gas emissions and a rise in global temperatures. To mitigate these effects, many countries are shifting to cleaner and more sustainable forms of energy, such as electric vehicles (EVs). The adoption of EVs presents a significant opportunity to reduce our dependence on oil, lower greenhouse gas emissions, and enhance energy security. This chapter will explore the potential of EVs in reducing our dependence on oil and the benefits of this transition for the environment.

EVs and Oil Dependence:

The transportation sector is a major consumer of oil, accounting for nearly two-thirds of global oil consumption. The dependence on oil makes countries vulnerable to price shocks, supply disruptions, and geopolitical tensions. The shift towards EVs can help to reduce this dependence on oil, as they do not rely on fossil fuels for power. EVs can be powered by renewable energy sources, such as solar and wind power, reducing the reliance on oil and improving energy security.

EVs can also help to reduce oil imports, particularly in countries that rely heavily on oil imports. In the United States, for example, oil imports accounted for 36% of petroleum consumption in 2020, according to the U.S. Energy Information Administration. This dependence on foreign oil can have economic and security implications. By shifting to EVs, countries can reduce their dependence on foreign oil and improve their energy security.

Environmental Benefits:

EVs also have significant environmental benefits. They produce fewer greenhouse gas emissions than traditional gasoline-powered vehicles, reducing the carbon footprint of the transportation sector. According to the International Energy Agency, the deployment of EVs could lead to a reduction of 1.3 gigatons of CO_2 emissions per year by 2030, which is equivalent to the annual emissions of Japan.

EVs also produce fewer pollutants that contribute to air pollution, such as nitrogen oxides and particulate matter. These pollutants have been linked to respiratory illnesses and can be harmful to human health. By reducing the emissions of these pollutants, the adoption of EVs can lead to improved air quality and public health.

Challenges:

Despite the potential benefits of EVs, there are still challenges that need to be addressed. One of the most significant challenges is range anxiety, which is the fear that an EV will run out of charge before reaching its destination. The limited range of EVs and the availability of charging infrastructure can make it difficult for drivers to travel long distances, particularly in rural areas.

The availability of charging infrastructure is also a concern. Many countries lack the necessary infrastructure to support the widespread adoption of EVs. This includes the availability of charging stations, as well as the capacity of the electricity grid to handle the increased demand for electricity. The deployment of charging infrastructure will require significant investment, coordination, and planning.

Another challenge is the cost of EVs. While the cost of batteries has decreased in recent years, EVs remain more expensive than gasoline-powered vehicles. This can make it difficult for some consumers to afford EVs, particularly in low-income countries. Governments and industry can help to address this challenge by providing incentives, such as tax credits or rebates, to encourage the adoption of EVs.

Conclusion:

The adoption of EVs presents a significant opportunity to reduce our dependence on oil and improve the environmental sustainability of the transportation sector. EVs produce fewer greenhouse gas emissions and pollutants than traditional gasoline-powered vehicles, reducing the carbon footprint of the sector and improving air quality. However, there are still challenges that need to be addressed, such as range anxiety, the availability of charging infrastructure, and the cost of EVs. Addressing these challenges will require significant investment, coordination, and planning from governments and industry.

Electric vehicles (EVs) have the potential to revolutionize the transportation industry, but their widespread adoption faces several challenges. These challenges range from technological barriers to societal factors that influence consumer behavior. In this chapter, we will discuss the key challenges to the widespread adoption of electric vehicles and explore possible solutions to overcome them.

1. Limited Range Anxiety

Range anxiety is one of the most significant barriers to the widespread adoption of electric vehicles. The limited range of EVs can be a major concern for consumers, especially for those who frequently drive long distances. While the range of EVs has improved in recent years, it still falls short of the range offered by traditional internal combustion engine (ICE) vehicles.

To address this challenge, automakers are developing EVs with longer ranges and faster charging times. Advances in battery technology have helped to increase the range of EVs significantly, and the emergence of fast-charging infrastructure has made it possible to charge EVs quickly on long trips. Moreover, policymakers are incentivizing the development of charging infrastructure to increase the accessibility and convenience of charging stations.

2. High Purchase Price

Another significant challenge to the widespread adoption of electric vehicles is their high purchase price. EVs are generally more expensive than their ICE counterparts, making them less accessible to a significant portion of the population. The high cost of batteries, which are a critical component of EVs, is a major contributor to the high purchase price.

To address this challenge, automakers are investing in battery production facilities to reduce the cost of batteries. The increasing demand for EVs is also driving down the cost of batteries. Additionally, policymakers are incentivizing the purchase of EVs through tax credits and rebates to reduce the overall cost of ownership.

3. Charging Infrastructure

The availability and accessibility of charging infrastructure is a critical factor influencing the widespread adoption of electric vehicles. EV owners need access to convenient and reliable charging stations to alleviate range anxiety and ensure that they can travel long distances without interruption.

To address this challenge, policymakers are investing in the development of charging infrastructure, including the construction of public charging stations and incentives for the installation of private charging stations. Automakers are also partnering with charging infrastructure providers to increase the availability of charging stations.

4. Consumer Awareness and Education

The lack of consumer awareness and education about electric vehicles is another challenge to their widespread adoption. Many consumers are unfamiliar with the technology and performance of electric vehicles, leading to skepticism about their practicality and reliability.

To address this challenge, policymakers and industry stakeholders are investing in education and awareness campaigns to inform consumers about the benefits of electric vehicles. Automakers are also investing in advertising and marketing campaigns to promote the adoption of EVs.

5. Battery Disposal

The disposal of batteries is another significant challenge to the widespread adoption of electric vehicles. EV batteries contain toxic materials that can harm the environment if not disposed of properly.

To address this challenge, policymakers are developing regulations to ensure the safe disposal and recycling of EV batteries. Automakers are also investing in research and development to improve the recyclability and sustainability of EV batteries.

Conclusion

The widespread adoption of electric vehicles is critical to reducing greenhouse gas emissions and addressing climate change. However, the adoption of EVs faces several challenges, ranging from technological barriers to societal factors that influence consumer behavior. To overcome these challenges, policymakers, industry stakeholders, and consumers must work together to invest in technological advancements, incentivize EV adoption, and increase consumer awareness and education. By doing so, we can ensure a cleaner and more sustainable future for generations to come.

Chapter 6: The Future of Electric Vehicles
The role of EVs in the transition to a clean energy economy

The world is facing a climate crisis that requires an urgent shift to a clean energy economy. The transportation sector is a major contributor to greenhouse gas emissions, accounting for around 16% of global emissions. As electric vehicles (EVs) become more accessible and affordable, they offer a promising solution to reduce the carbon footprint of transportation.

EVs have the potential to play a key role in the transition to a clean energy economy in several ways. First, they offer a direct way to reduce greenhouse gas emissions. EVs produce zero emissions at the tailpipe, meaning they emit no pollutants or greenhouse gases. This is a stark contrast to gasoline-powered vehicles, which emit significant amounts of carbon dioxide and other pollutants.

Second, EVs can contribute to the integration of renewable energy into the electricity grid. As renewable energy sources like wind and solar become more prevalent, they can create challenges for grid operators due to their intermittent nature. EVs can be used as a means of energy storage, helping to balance the grid and reduce the need for fossil fuel-based peaker plants.

Third, EVs can support the growth of domestic industries and jobs. As the demand for EVs increases, it can drive the growth of domestic manufacturing, battery production, and charging infrastructure. This can create new jobs and boost economic growth in regions that invest in EV technology.

However, there are several challenges that must be addressed to fully realize the potential of EVs in the transition to a clean energy economy.

One challenge is the need for a comprehensive charging infrastructure. EVs require reliable access to charging stations to enable long-distance travel and ease of use. Governments and private entities must work together to invest in charging infrastructure and ensure that it is accessible to all EV drivers.

Another challenge is the need for affordable and accessible EVs. While the cost of EVs has been decreasing in recent years, they remain more expensive than gasoline-powered vehicles for many consumers. Governments can support the growth of the EV market by offering incentives and subsidies to reduce the cost of EVs and increase their accessibility.

Furthermore, the transition to a clean energy economy will require significant changes in energy policies and regulations. Governments must implement policies that promote the use of renewable energy sources and discourage the use of fossil fuels. This will require coordination among governments at all levels and the engagement of the private sector and civil society.

Finally, the growth of the EV market will require innovation and collaboration among manufacturers and other stakeholders. As the technology evolves, manufacturers must continue to improve the performance and affordability of EVs. Collaboration among stakeholders will be critical to ensure that charging infrastructure is built out in a coordinated and efficient manner.

In conclusion, EVs offer a promising solution to reduce greenhouse gas emissions from the transportation sector and support the transition to a clean energy economy. However, significant challenges remain that must be addressed through coordinated efforts by governments, the private sector, and civil society. By working together, we can unlock the full potential of EVs and create a more sustainable future for all.

The potential for new battery technology

As the electric vehicle (EV) market continues to grow, advancements in battery technology have become a key focus for manufacturers and researchers alike. While lithium-ion batteries have been the primary energy storage solution for EVs for several years, there is growing interest in developing new battery technologies that could offer increased energy density, faster charging times, longer lifespans, and lower costs.

One of the most promising battery technologies currently being researched is the solid-state battery. Solid-state batteries use a solid electrolyte instead of a liquid one, which could provide several advantages over traditional lithium-ion batteries. Solid-state batteries have the potential to offer higher energy density, faster charging times, longer lifespans, and improved safety due to their reduced risk of leaking or catching fire.

Another emerging battery technology is the lithium-sulfur battery. Lithium-sulfur batteries have a higher theoretical energy density than lithium-ion batteries, which means they could potentially store more energy in the same amount of space. However, lithium-sulfur batteries have some challenges to overcome, including a shorter lifespan and difficulty maintaining their energy density over multiple charge cycles.

Researchers are also exploring the potential of using other materials in batteries, such as sodium, magnesium, and zinc. Sodium-ion batteries, for example, are being studied as a potential alternative to lithium-ion batteries. Sodium is more abundant and less expensive than lithium, which could make sodium-ion batteries a more sustainable and cost-effective option in the long run. Magnesium and zinc batteries are also being researched for their potential to offer higher energy density and longer lifespans than lithium-ion batteries.

In addition to developing new battery technologies, researchers are also exploring ways to improve the performance of lithium-ion batteries. One approach is to use new electrode materials, such as silicon, which has a higher energy density than graphite, the most commonly used electrode material in lithium-ion batteries. Other approaches include improving the structure of the battery electrodes, using new electrolyte materials, and developing better battery management systems to optimize charging and discharging.

While there is still much work to be done to develop and commercialize these new battery technologies, their potential benefits for the EV industry are significant. By offering increased energy density, faster charging times, longer lifespans, and lower costs, new battery technologies could help to accelerate the transition to electric transportation and reduce our reliance on fossil fuels. As such, the development of new battery technologies is a crucial area of research in the EV industry and one that is likely to have a major impact on the future of transportation.

The impact of government policies and regulations

The impact of government policies and regulations on the future of electric vehicles (EVs) is a complex and multifaceted issue. Government policies and regulations can have a significant impact on the development, adoption, and success of EVs.

One of the most significant ways that government policies can impact the future of EVs is through financial incentives. Many governments around the world offer subsidies and tax credits to encourage the purchase of EVs. These incentives can significantly reduce the upfront cost of EVs, making them more affordable and attractive to consumers. Additionally, some governments offer rebates or other incentives to businesses that invest in EV infrastructure, such as charging stations.

In the United States, for example, the federal government offers a tax credit of up to $7,500 for the purchase of a new electric vehicle. Many states also offer additional incentives, such as tax credits, rebates, or free access to high-occupancy vehicle lanes. Similar programs exist in countries around the world, such as Norway's zero-emissions vehicle (ZEV) incentive program, which offers significant tax breaks for electric vehicle buyers.

Another way that government policies can impact the future of EVs is through regulations aimed at reducing greenhouse gas emissions. Many governments have set ambitious goals for reducing emissions and have implemented regulations that require automakers to produce vehicles with lower emissions. These regulations can encourage automakers to develop and produce more EVs, as they often have lower emissions than traditional gasoline or diesel vehicles.

In Europe, for example, the European Union has set emissions standards that require automakers to reduce emissions from new vehicles sold in the EU. This has led many automakers to invest heavily in the development of electric and hybrid vehicles to meet these standards.

In addition to financial incentives and emissions regulations, government policies can also impact the future of EVs through investments in infrastructure. EVs require charging stations, and many governments have made significant investments in the development of charging infrastructure to support the growth of EVs. These investments can make it more convenient for consumers to own and operate EVs, reducing one of the major barriers to adoption.

For example, the Chinese government has made significant investments in charging infrastructure, with plans to install 4.8 million charging stations by 2025. In the United States, the federal government has provided funding for the development of a national network of electric vehicle charging stations through the EV Everywhere Grand Challenge.

However, government policies and regulations can also create challenges for the growth and adoption of EVs. For example, in some countries, automakers are required to produce a certain percentage of zero-emissions vehicles as part of their overall fleet, known as a zero-emissions vehicle mandate. While this can encourage the development and production of EVs, it can also create a significant burden for automakers that are unable to meet these requirements.

Additionally, some governments have implemented policies that create uncertainty for automakers and consumers. For example, changes to incentives or regulations can create uncertainty about the future of EVs, which can make it more difficult for automakers to plan for future production and for consumers to make informed decisions about purchasing EVs.

In conclusion, government policies and regulations can have a significant impact on the future of electric vehicles. Financial incentives, emissions regulations, and investments in infrastructure can all encourage the growth and adoption of EVs. However, policies that create uncertainty or place burdens on automakers can create challenges for the development and success of EVs. As such, policymakers must carefully consider the impact of their policies on the future of EVs and work to create a supportive regulatory environment that encourages the growth and adoption of EVs as part of a transition to a cleaner energy future.

Chapter 7: The Challenges and Opportunities Ahead
The need for charging infrastructure

Introduction: As electric vehicles (EVs) continue to gain popularity and become more mainstream, the need for charging infrastructure has become increasingly apparent. In this chapter, we will discuss the importance of charging infrastructure for EVs, the current state of charging infrastructure, and the challenges and opportunities associated with building out charging networks.

Importance of Charging Infrastructure: One of the biggest barriers to widespread adoption of EVs is range anxiety, or the fear that an EV will run out of charge before reaching its destination. This fear can be alleviated by building out a robust charging infrastructure that enables EV drivers to charge their vehicles conveniently and reliably. A reliable charging infrastructure will increase consumer confidence in EVs and reduce range anxiety, making it more likely that consumers will consider an EV as a viable option.

Current State of Charging Infrastructure: The current state of charging infrastructure varies widely depending on the region and country. In some areas, such as California and Europe, there are already extensive charging networks that provide convenient access to EV charging. In other areas, however, charging infrastructure is scarce, making it difficult for EV drivers to travel long distances.

In the United States, the government has taken steps to encourage the development of charging infrastructure. For example, the Department of Energy has launched the EV Everywhere Grand Challenge, which aims to make EVs as affordable and convenient as gasoline-powered vehicles within a decade. The challenge includes a goal of deploying charging infrastructure that enables EV drivers to travel coast-to-coast on electric power alone by 2022.

Challenges and Opportunities: Building out a robust charging infrastructure is not without its challenges. One of the biggest challenges is the cost of building and maintaining charging stations. Charging infrastructure can be expensive to install, and there is currently no clear business model for charging station operators. In addition, there is a lack of standardization in the charging industry, which can make it difficult for charging stations to work with different EV models.

Despite these challenges, there are also opportunities associated with building out charging infrastructure. For example, there is a growing market for EV charging services, and companies that build out charging networks can potentially profit from this market. In addition, the development of new battery technology could lead to faster charging times and longer ranges, making EVs more practical for long-distance travel.

Conclusion: In conclusion, the need for charging infrastructure is critical to the success of EVs. A reliable charging infrastructure will alleviate range anxiety and increase consumer confidence in EVs. Although there are challenges associated with building out charging networks, there are also opportunities for companies that invest in charging infrastructure. As the EV market continues to grow, it is likely that the demand for charging infrastructure will only increase, making it a critical component of the EV ecosystem.

The impact on the electricity grid

As electric vehicle (EV) adoption continues to grow, concerns about the impact on the electricity grid have become more prominent. The increased demand for electricity to charge EVs has raised questions about grid capacity, stability, and reliability. However, with proper planning and investment, the impact of EVs on the grid can be minimized, and there are also opportunities for EVs to support the development of a more flexible and sustainable electricity system.

Grid Capacity

One of the primary concerns about the impact of EVs on the grid is whether the existing infrastructure can handle the increased demand for electricity. According to the International Energy Agency (IEA), the number of EVs on the road is expected to reach 145 million by 2030, which could require an additional 640 TWh of electricity globally. This represents a significant increase in demand, but it is important to note that the growth will not be uniform across all regions and that the rate of EV adoption will depend on a variety of factors, including government policies, incentives, and consumer preferences.

In some regions, the existing grid infrastructure may be sufficient to handle the additional demand for electricity from EVs, while in others, upgrades may be necessary. Utilities and grid operators will need to assess their existing infrastructure and plan for upgrades as needed to ensure that the grid can handle the additional load. This may involve upgrades to transmission and distribution lines, transformers, and substation equipment.

Stability and Reliability

Another concern related to the impact of EVs on the grid is the potential impact on grid stability and reliability. The increased demand for electricity from EVs can create fluctuations in demand that could affect grid stability. For example, if a large number of EVs are charging at the same time in a particular area, this could create a local surge in demand that could strain the grid.

To address this concern, utilities and grid operators are exploring a variety of solutions, including smart charging and vehicle-to-grid (V2G) technology. Smart charging allows EVs to be charged during off-peak hours when electricity demand is lower, which can help to balance the load on the grid. V2G technology allows EVs to not only draw power from the grid but also to feed power back into the grid, which can provide additional grid stability and flexibility.

Opportunities for Grid Integration

Despite the challenges associated with the impact of EVs on the grid, there are also opportunities for EVs to support the development of a more flexible and sustainable electricity system. EVs can provide a source of flexible demand that can be managed to support grid stability and integration of variable renewable energy sources such as wind and solar. EVs can also support the development of local energy systems, such as microgrids, which can improve grid resilience and provide backup power in case of outages.

In addition, EVs can also play a role in providing ancillary services to the grid, such as frequency regulation and voltage control. These services are critical for maintaining grid stability and reliability and are typically provided by conventional power plants. However, as more renewable energy sources are integrated into the grid, the need for these services may increase, and EVs could help to fill this gap.

Conclusion

The impact of EVs on the electricity grid is a complex and evolving issue that will require careful planning and investment by utilities, grid operators, and governments. However, with the right policies and investments in place, the impact of EVs on the grid can be minimized, and there are also opportunities for EVs to support the development of a more flexible and sustainable electricity system. As the adoption of EVs continues to grow, it will be important to monitor the impact on the grid and to develop solutions that ensure the reliability and stability of the electricity system while maximizing the benefits of EVs.

The potential for electric autonomous vehicles

Electric autonomous vehicles, or self-driving cars, are rapidly becoming a reality. These vehicles use a variety of sensors and technologies to perceive their surroundings and make decisions based on that information. This technology has the potential to revolutionize transportation by reducing accidents and fatalities, improving traffic flow, and reducing the need for personal car ownership. When combined with electric power, these benefits can be amplified even further.

One of the most significant advantages of electric autonomous vehicles is the potential to dramatically reduce accidents and fatalities. According to the National Highway Traffic Safety Administration (NHTSA), 94% of serious crashes are caused by human error. Self-driving cars remove this element of risk, as they are programmed to obey traffic laws, avoid collisions, and make safe decisions. This could lead to a significant reduction in the number of accidents and fatalities on the roads.

Another benefit of electric autonomous vehicles is the potential to improve traffic flow. These cars are equipped with advanced sensors that allow them to communicate with one another and adjust their speed and direction accordingly. This can reduce congestion, as well as the need for stoplights and other traffic control devices. Additionally, self-driving cars can drive closer together, reducing the space between vehicles and allowing more cars to fit on the road.

Electric autonomous vehicles can also reduce the need for personal car ownership. With the advent of ride-sharing services like Uber and Lyft, many people are already forgoing car ownership in favor of on-demand transportation. Self-driving cars could take this a step further by providing a safe, reliable, and affordable alternative to traditional car ownership. This could lead to fewer cars on the road, reduced traffic congestion, and a lower carbon footprint.

There are, however, several challenges to the widespread adoption of electric autonomous vehicles. One major challenge is the development of reliable and affordable technology. While companies like Tesla, Google, and Uber have made significant progress in this area, there is still a long way to go before fully autonomous vehicles are a common sight on our roads.

Another challenge is the need for new infrastructure to support self-driving cars. This includes charging stations for electric cars, as well as communication networks that allow self-driving cars to communicate with one another and with traffic control systems. Additionally, there may be legal and regulatory challenges to overcome, as governments work to develop new laws and regulations to govern the use of self-driving cars.

Despite these challenges, the potential benefits of electric autonomous vehicles are too significant to ignore. With their ability to reduce accidents, improve traffic flow, and reduce the need for personal car ownership, these cars could have a profound impact on the way we live and work. As technology continues to improve and infrastructure is developed to support them, electric autonomous vehicles are likely to become a common sight on our roads in the years to come.

Conclusion
The future of electric vehicles and the transportation industry

The future of electric vehicles is an exciting topic, with many possibilities for growth and innovation. As we've seen throughout this book, electric vehicles have come a long way since their early days, with major advancements in battery technology and increased public awareness driving their widespread adoption. In this concluding chapter, we will discuss the potential for electric vehicles to continue transforming the transportation industry in the years to come.

One of the most significant drivers of change in the transportation industry is the need to reduce greenhouse gas emissions and combat climate change. Electric vehicles are a key solution to this challenge, as they emit significantly less carbon dioxide and other pollutants than traditional gasoline-powered vehicles. As governments around the world continue to push for stricter emissions standards and incentives for electric vehicle adoption, we can expect the demand for electric vehicles to increase.

Another trend driving the growth of electric vehicles is the continued development of autonomous driving technology. While electric vehicles and autonomous driving are not inherently linked, many see them as complementary technologies that can work together to create a more sustainable and efficient transportation system. By eliminating the need for human drivers, electric autonomous vehicles could reduce traffic congestion, improve safety, and make it easier for people to get around without owning a car.

However, there are also challenges ahead for electric vehicles. One of the most significant is the need for charging infrastructure. While many countries have made significant progress in building out charging networks, there is still a long way to go before electric vehicles can truly rival gasoline-powered cars in terms of convenience and accessibility. This will require significant investment in charging infrastructure, as well as new technologies that allow for faster charging and longer range.

Another challenge is the impact that electric vehicles could have on the electricity grid. As the number of electric vehicles on the road grows, there will be increased demand for electricity to power them. This could put significant strain on the grid, particularly if charging is concentrated during peak demand periods. Addressing this challenge will require a coordinated effort between utilities, governments, and electric vehicle manufacturers to ensure that the grid can accommodate the increased demand for electricity.

Despite these challenges, the future of electric vehicles looks bright. As battery technology continues to improve, we can expect electric vehicles to become more affordable and accessible to a wider range of consumers. We can also expect to see new players enter the market, including startups and established companies from outside the automotive industry. As electric vehicles become more mainstream, they will likely become a key driver of innovation and growth in the transportation industry.

In conclusion, electric vehicles have come a long way since their early days, and they are poised to continue transforming the transportation industry in the years to come. While there are challenges ahead, including the need for charging infrastructure and managing the impact on the electricity grid, the potential benefits of electric vehicles are significant. As we look to the future, we can expect electric vehicles to play an increasingly important role in creating a more sustainable, efficient, and connected transportation system.

The potential for electric vehicles to revolutionize the way we travel

The widespread adoption of electric vehicles has the potential to revolutionize the way we travel, and it represents a significant shift in the transportation industry. The benefits of electric vehicles are numerous, from their reduced environmental impact to their lower operating costs, but they also present new challenges that must be addressed for their widespread adoption. In this section, we will discuss the potential for electric vehicles to revolutionize the way we travel and the challenges that must be addressed for this to happen.

One of the most significant ways that electric vehicles have the potential to revolutionize the way we travel is by reducing our dependence on fossil fuels. The transportation sector is one of the largest contributors to greenhouse gas emissions, and electric vehicles have the potential to significantly reduce these emissions. By replacing gasoline and diesel-powered vehicles with electric vehicles, we can reduce our dependence on oil and the associated geopolitical issues that come with it.

In addition to their environmental benefits, electric vehicles also offer new opportunities for the transportation industry. One of the most significant opportunities is the potential for electric vehicles to be integrated into a smart grid. With the increasing adoption of renewable energy sources such as wind and solar power, the integration of electric vehicles into the grid can help to balance the supply and demand of electricity. For example, electric vehicles could be charged during times of low demand, such as at night, and then discharge their energy back into the grid during times of high demand, such as during hot summer afternoons.

Another significant opportunity that electric vehicles present is the potential for new business models and revenue streams. As electric vehicles become more prevalent, new companies and industries will emerge to support their adoption. For example, electric vehicle charging infrastructure providers, battery manufacturers, and electric vehicle component suppliers will all see increased demand for their products and services.

However, there are also significant challenges that must be addressed for electric vehicles to reach their full potential. One of the most significant challenges is the need for infrastructure to support electric vehicle charging. While the number of charging stations is increasing, there is still a significant lack of charging infrastructure in many areas. This lack of infrastructure can be a barrier to the widespread adoption of electric vehicles, as it can limit the range and convenience of electric vehicles.

Another challenge that must be addressed is the cost of electric vehicles. While the cost of electric vehicles has decreased significantly in recent years, they are still more expensive than traditional gasoline-powered vehicles. This cost differential can be a significant barrier to adoption, particularly for those on a tight budget.

Finally, there is also a need for continued investment in battery technology to improve the range and performance of electric vehicles. While the range of electric vehicles has increased significantly in recent years, it is still a concern for many consumers. Additionally, as the adoption of electric vehicles increases, there is a need for increased battery production capacity to meet demand.

In conclusion, the potential for electric vehicles to revolutionize the way we travel is significant. They offer numerous benefits, from their reduced environmental impact to their potential for new business models and revenue streams. However, there are also significant challenges that must be addressed for their widespread adoption. These challenges include the need for infrastructure to support electric vehicle charging, the cost of electric vehicles, and the need for continued investment in battery technology. With continued investment and support, electric vehicles have the potential to transform the transportation industry and provide a cleaner, more sustainable future.

The challenges and opportunities ahead for the EV industry

The electric vehicle (EV) industry has come a long way since the early days of limited range and availability. The advancements in battery technology and government incentives have driven the growth of the EV market, with more and more people choosing to make the switch to electric vehicles. However, as the industry continues to grow, it faces a number of challenges and opportunities that will shape the future of transportation.

One of the biggest challenges facing the EV industry is the need for charging infrastructure. While more and more charging stations are being installed, there is still a long way to go to meet the demand of the growing number of electric vehicles on the road. The lack of charging infrastructure has been cited as a major barrier to widespread adoption of EVs, with concerns over "range anxiety" – the fear of running out of power before reaching a charging station. To overcome this challenge, government and private sector investment in charging infrastructure will be crucial.

Another challenge facing the EV industry is the impact on the electricity grid. As more EVs are adopted, there will be an increase in demand for electricity, which will require upgrades to the grid to ensure that it can handle the increased load. This presents an opportunity for the integration of renewable energy sources, such as wind and solar, into the grid, which can help reduce carbon emissions and increase the sustainability of the transportation sector.

The potential for electric autonomous vehicles also presents both challenges and opportunities for the EV industry. Autonomous vehicles have the potential to revolutionize transportation, providing safer and more efficient travel. However, they also require significant investment in research and development, as well as the infrastructure to support them. Additionally, there are concerns over cybersecurity and data privacy, which will need to be addressed to ensure the safety and security of passengers.

Despite these challenges, the EV industry presents numerous opportunities for growth and innovation. As the technology continues to improve, electric vehicles will become more affordable and accessible, making them an attractive option for a wider range of consumers. In addition, the shift towards electrification will create new opportunities for job creation and economic growth, particularly in the renewable energy sector.

The EV industry also has the potential to revolutionize the way we think about transportation. By reducing our dependence on fossil fuels, we can help mitigate the impacts of climate change and create a more sustainable future. The development of a clean energy transportation sector will also reduce our dependence on foreign oil, providing energy security and reducing the geopolitical tensions associated with oil production and distribution.

In conclusion, the challenges and opportunities facing the EV industry are significant, but the potential for growth and innovation is equally large. With continued investment in charging infrastructure, grid upgrades, and research and development, the EV industry can continue to drive progress towards a clean energy future. The adoption of electric vehicles can not only improve air quality and mitigate the impacts of climate change, but also revolutionize the way we travel and live our lives. The future of transportation is electric, and the EV industry is poised to lead the way.

THE END

Key Terms and Definitions

To help you better understand the language and concepts related to aging and older adults, below you will find a list of key terms and their definitions.

1. Electric vehicle (EV): A vehicle that is powered by an electric motor and a battery rather than an internal combustion engine that runs on gasoline or diesel fuel.

2. Battery electric vehicle (BEV): A type of electric vehicle that is powered solely by an electric motor and a battery, with no backup gasoline engine.

3. Plug-in hybrid electric vehicle (PHEV): A type of electric vehicle that has both an electric motor and a backup gasoline engine, and can be recharged by plugging it into an external power source.

4. Range anxiety: The fear or anxiety that a driver of an electric vehicle may experience when they are concerned about running out of battery charge before reaching their destination.

5. Charging infrastructure: The network of charging stations and other facilities needed to support the charging of electric vehicles.

6. Lithium-ion battery: A type of rechargeable battery used in electric vehicles that is known for its high energy density and long cycle life.

7. Kilowatt-hour (kWh): A unit of energy used to measure the amount of electricity stored in a battery.

8. Level 1 charging: The slowest type of electric vehicle charging, typically done using a standard 120-volt household outlet.

9. Level 2 charging: A faster type of electric vehicle charging that uses a 240-volt circuit, often found in commercial charging stations or installed at home.

10. Direct current (DC) fast charging: The fastest type of electric vehicle charging, which uses a high-powered DC charging station to provide a rapid charge to the vehicle's battery.

Supporting Materials

Introduction

- U.S. Department of Energy. (2021). Electric Vehicles 101.
https://www.energy.gov/eere/electricvehicles/electric-
vehicles-101

Chapter 1

- Kirsch, D. (2000). The Electric Vehicle and the Burden of
History. Rutgers University Press.

- National Renewable Energy Laboratory. (2021). History of
Electric Vehicles.

https://www.nrel.gov/transportation/electric-
vehicles/history.html

Chapter 2

- Kirsch, D. (2000). The Electric Vehicle and the Burden of
History. Rutgers University Press.

- Sperling, D., & Gordon, D. (2018). Two Billion Cars:
Driving Toward Sustainability. Oxford University Press.

Chapter 3

- Shnayerson, M. (2018). The Car That Could: The Inside
Story of GM's Revolutionary Electric Vehicle. Random
House.

- Vance, A. (2015). Elon Musk: Tesla, SpaceX, and the Quest
for a Fantastic Future. Ecco.

Chapter 4

- International Energy Agency. (2020). Global EV Outlook 2020: Entering the Decade of Electric Drive? https://www.iea.org/reports/global-ev-outlook-2020

- Jaffe, A. B., & Stavins, R. N. (2020). The long and winding road to electric cars: A cost–benefit analysis. Journal of Economic Perspectives, 34(2), 92-114.

Chapter 5

- Delucchi, M. A., & Jacobson, M. Z. (2011). Providing all global energy with wind, water, and solar power, Part I: Technologies, energy resources, quantities and areas of infrastructure, and materials. Energy Policy, 39(3), 1154-1169.

- National Renewable Energy Laboratory. (2021). Environmental Benefits of Electric Vehicles. https://www.nrel.gov/transportation/electric-vehicles/benefits.html

Chapter 6

- International Energy Agency. (2020). Global EV Outlook 2020: Entering the Decade of Electric Drive? https://www.iea.org/reports/global-ev-outlook-2020

- Sripad, S. (2020). Electric vehicle adoption and the future of the auto industry. Journal of the Association of Environmental and Resource Economists, 7(3), 519-558.

Chapter 7

- National Renewable Energy Laboratory. (2021). Charging Infrastructure Research. https://www.nrel.gov/transportation/electric-vehicle-charging-infrastructure.html
- The White House. (2021). Fact Sheet: Biden Administration Jumpstarts Offshore Wind Energy Projects to Create Jobs. https://www.whitehouse.gov/briefing-room/statements-releases/2021/03/29/fact-sheet-biden-administration-jumpstarts-offshore-wind-energy-projects-to-create-jobs/
Conclusion:
- International Energy Agency. (2020). Global EV Outlook 2020: Entering the Decade of Electric Drive? https://www.iea.org/reports/global-ev-outlook-2020
- Sripad, S. (2020). Electric vehicle adoption and the future of the auto industry. Journal of the Association of Environmental and Resource Economists, 7(3), 519-558.